HUDSON LIBRARY & HISTORICAL SOCIETY

D0744411

by Andreu Llamas
Illustrated by Gabriel Casadevall and Ali Garousi

Gareth Stevens Publishing
MILWAUKEE

**For a free color catalog describing Gareth Stevens' list of high-quality books and multimedia programs, call 1-800-542-2595 (USA) or 1-800-461-9120 (Canada). Gareth Stevens Publishing's Fax: (414) 225-0377.
See our catalog, too, on the World Wide Web: http://gsinc.com**

The editor would like to extend special thanks to Jan W. Rafert, Curator of Primates and Small Mammals, Milwaukee County Zoo, Milwaukee, Wisconsin, for his kind and professional help with the information in this book.

Library of Congress Cataloging-in-Publication Data

Llamas, Andreu.
[Esponja. English]
Sponges: filters of the sea / by Andreu Llamas ; illustrated by Gabriel Casadevall and Ali Garousi.
p. cm. – (Secrets of the animal world)
Includes bibliographical references and index.
Summary: Examines the different types, habitat, physical characteristics, enemies, and ancestors of sponges.
ISBN 0-8368-1645-5 (lib. bdg.)
1. Sponges–Juvenile literature. [1. Sponges.] I. Casadevall, Gabriel, ill. II. Garousi, Ali, ill. III. Title. IV. Series.
QL371.6.L5313 1997
593.4–dc21 97-8485

This North American edition first published in 1997 by
Gareth Stevens Publishing
1555 North RiverCenter Drive, Suite 201
Milwaukee, Wisconsin 53212 USA

Series editor: Patricia Lantier-Sampon
Editorial assistants: Diane Laska, Rita Reitci

Printed in the United States of America

1 2 3 4 5 6 7 8 9 01 00 99 98 97

CONTENTS

THE WORLD OF THE SPONGE

Where sponges live

Sponges are primitive, unusual animal filters that live in the sea. Most sponges are marine species (more than 80 percent), but others live in the fresh waters of rivers and lakes.

Sponges are found at all depths from the Equator to the poles, although they normally stay in shallow water. They can live stuck to the sea floor or to submerged objects along the coast. Some species also survive in the deep, dark abysses of the sea.

Calyx sponges are shaped like cups. They live on hard bases between 15 to 165 feet (5 to 50 meters) deep.

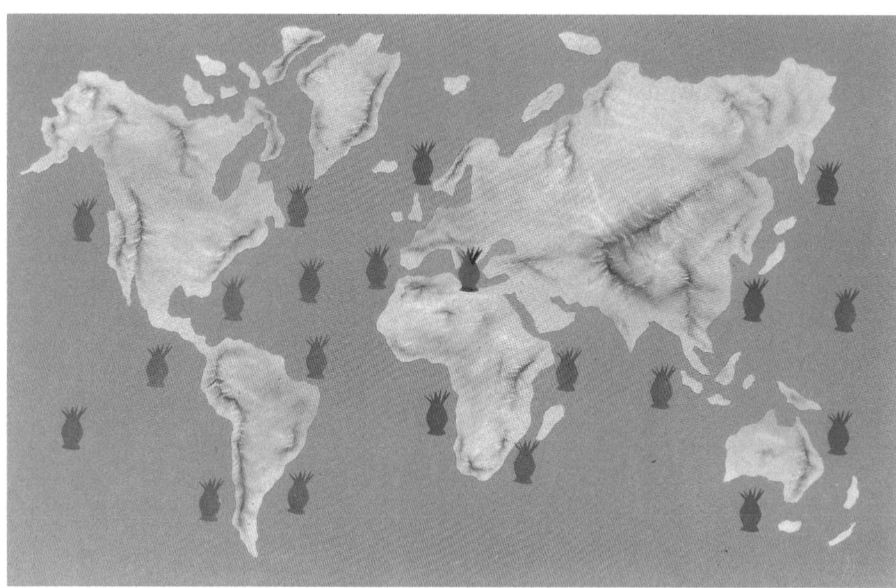

Sponges can be found from a few inches (cm) down to about 9,850 feet (3,000 m) deep.

Filtering the sea

Sponges spend their adult lives filtering sea water. Their bodies are covered with pore cells. Sponges eat nutritious particles filtered from the waters that pass through them. Breathing and the expulsion of wastes also occur through this filtering process. Adult sponges do not move around. Scientists once thought the sponges were vegetables.

Adult sponges have a great many pore cells through which the sea flows.

Sponges are animals that have very unusual shapes.

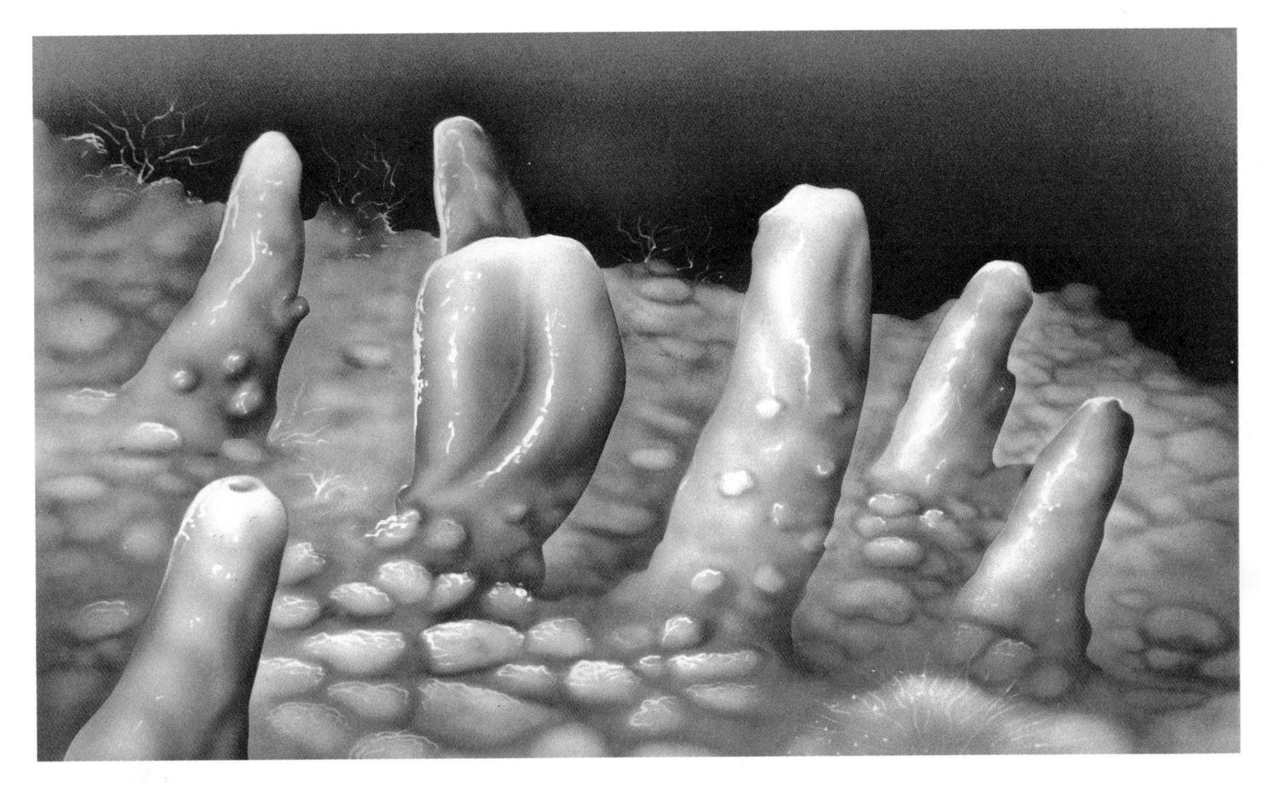

Classes of sponges

There are at least ten thousand sponge species classified according to three types: (1) the asconoids. These sponges have the most simple structures. They are usually less than 4 inches (10 cm) tall and shaped like a sack or tube with a slender body wall that surrounds a large cavity called the atrial cavity. Its exterior opening is called the osculum. The outside wall of the sponge is covered with pore cells through which water enters.

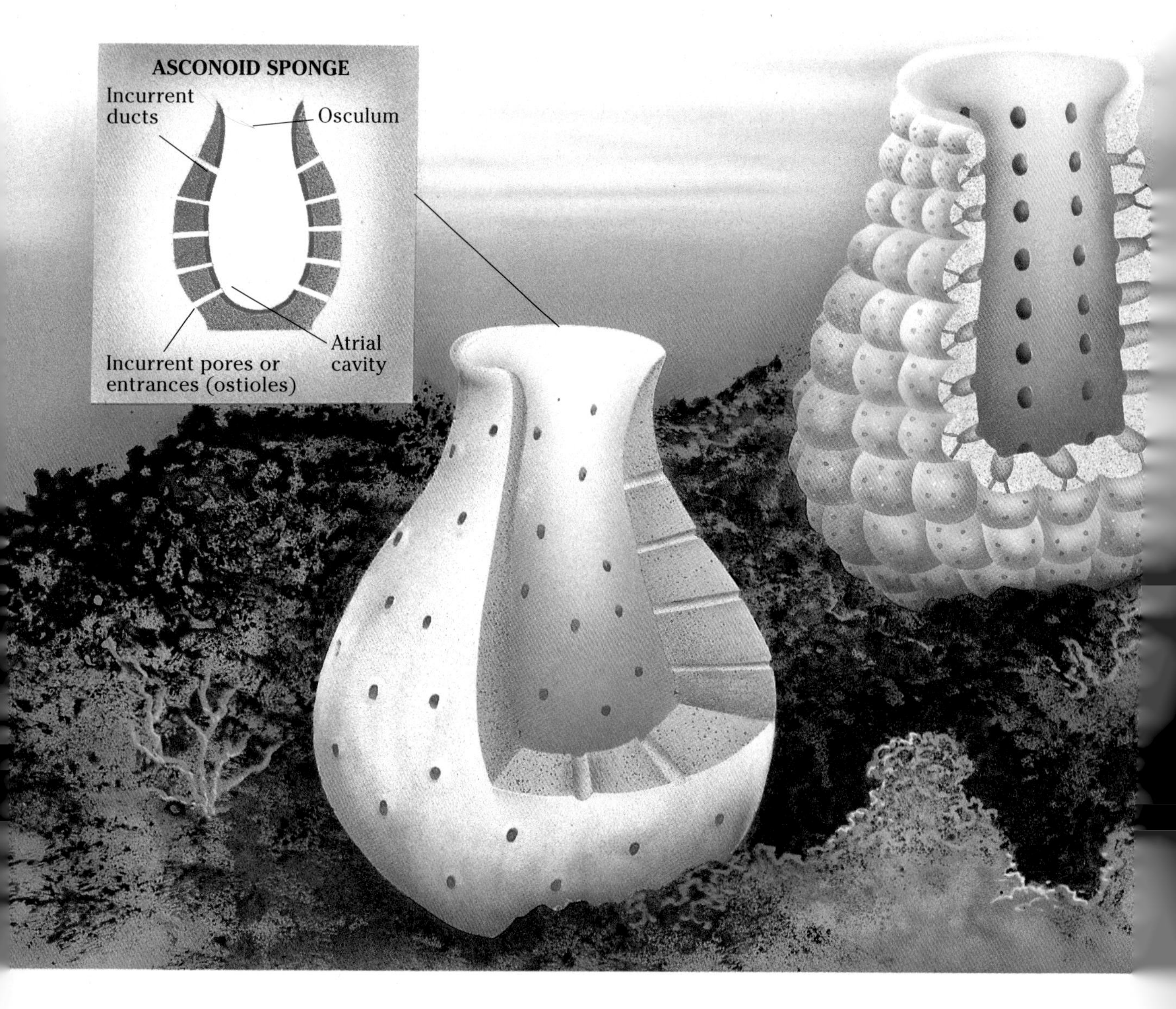

(2) the syconoids. These sponges are larger than the asconoids. They are tubular-shaped like the others and also have the same type of opening to the outside — the osculum. But the wall of this sponge is thicker and more complex. The smooth exterior has a series of evenly spaced circular holes. (3) the leuconoids. These sponges can be found alone or in groups. These are the largest sponges with the most effective filters. They have perfected the method of retaining nutritive particles.

INSIDE THE SPONGE

Sponge shapes do not have a specific symmetry. They grow in such various forms it is easy to think of them as vegetables instead of animals. Many different-looking sponges exist within the same species. Their appearance depends mainly on the environment. Sponges also come in a variety of colors.

WATER SYSTEM
Consists of a series of ducts and holes of differing sizes that take in, circulate, and expel water for the sponge.

FLAGELLATED CHAMBERS
Chambers lined with choanae, which capture food.

INCURRENT PORES
Water goes into the sponge through cells with tiny holes, called incurrent pores, or ostioles. These either cover the entire outer surface or only certain areas.

INCURRENT DUCTS

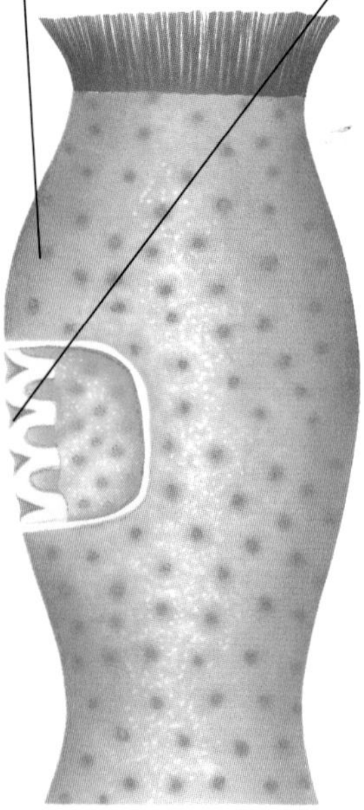

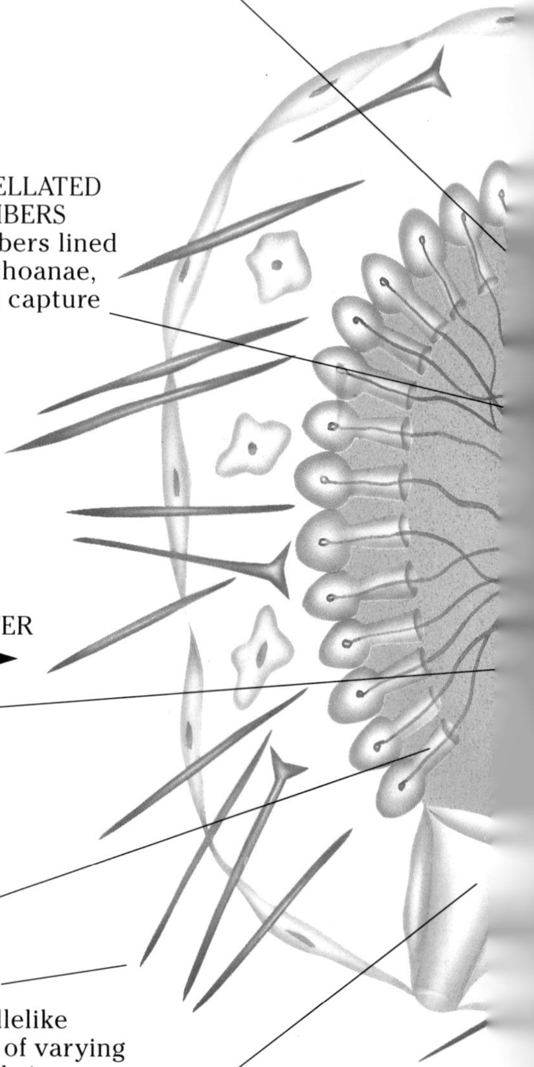

INCOMING WATER

EFFERENT DUCTS
A system of widening ducts that moves the water toward the outside.

FLAGELLATED CELLS
Choanae are cells with a flagellum that acts as a whip to capture food for the sponge. Choanae also move water out of the animal.

SPICULES
Hard, needlelike structures of varying shapes and sizes as small as 0.04 inch (1 mm). Spicules strengthen the animal's walls and help shape the sponge.

POROCYTES
Cylindrical cells, each with a central opening that forms a hole in the sponge's wall.

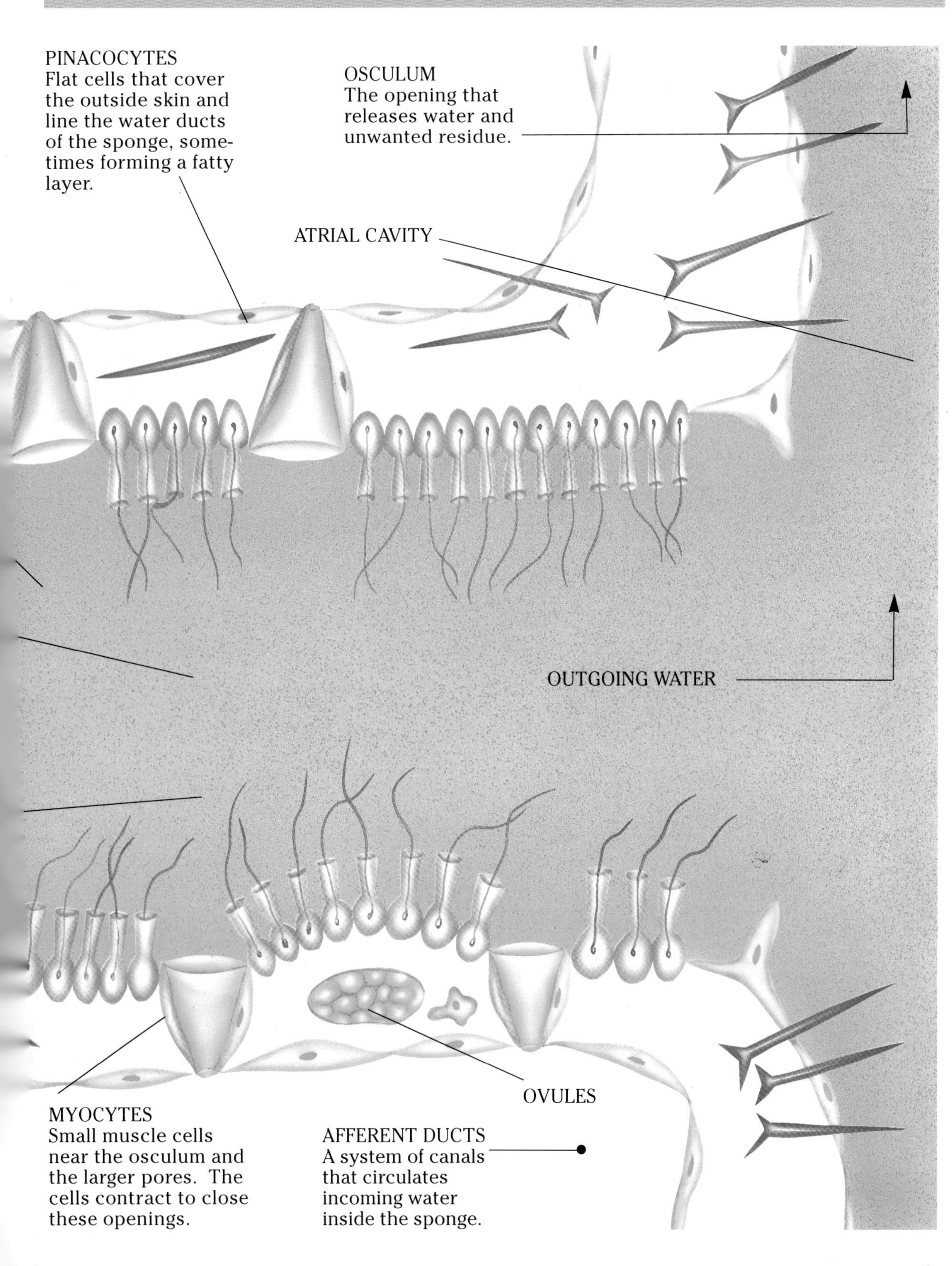
PINACOCYTES
Flat cells that cover the outside skin and line the water ducts of the sponge, sometimes forming a fatty layer.
OSCULUM
The opening that releases water and unwanted residue.
ATRIAL CAVITY
OUTGOING WATER
MYOCYTES
Small muscle cells near the osculum and the larger pores. The cells contract to close these openings.
OVULES
AFFERENT DUCTS
A system of canals that circulates incoming water inside the sponge.

FILTERS BY THE MILLIONS

Filtration system

The sponge retains all the nourishing particles that circulate inside its body. Through a system of water circulation, the sponge obtains the oxygen it needs to breathe and gets rid of unwanted residue. The sponge has an enormous pumping mechanism that pulls water into the interior and expels it after filtering.

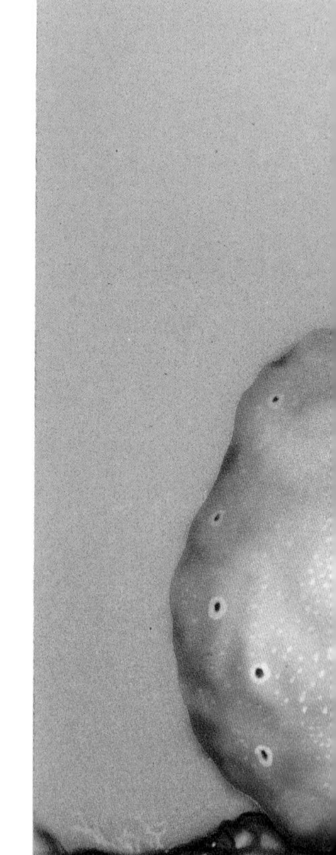

The pores that take water in are easy to see in all sponges.

The water route inside a syconoid sponge. Notice how the flagellae intercept particles of food.

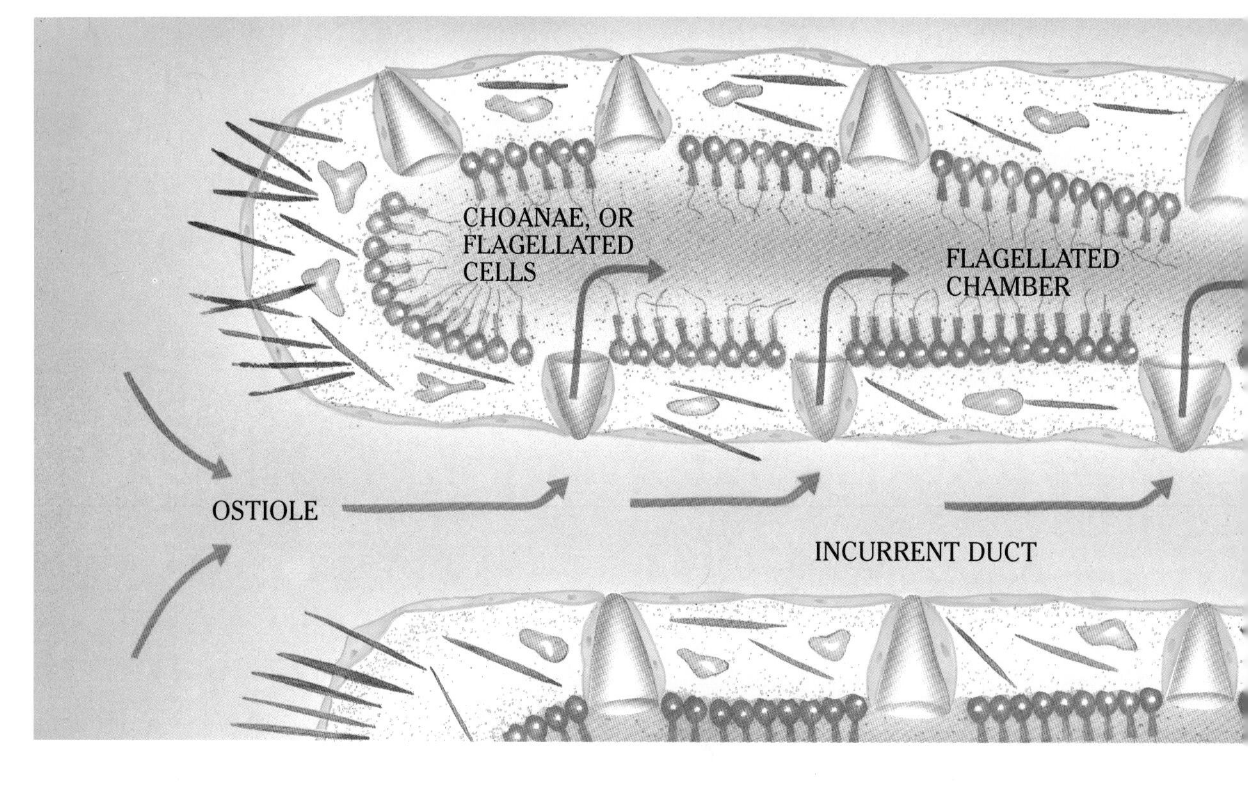

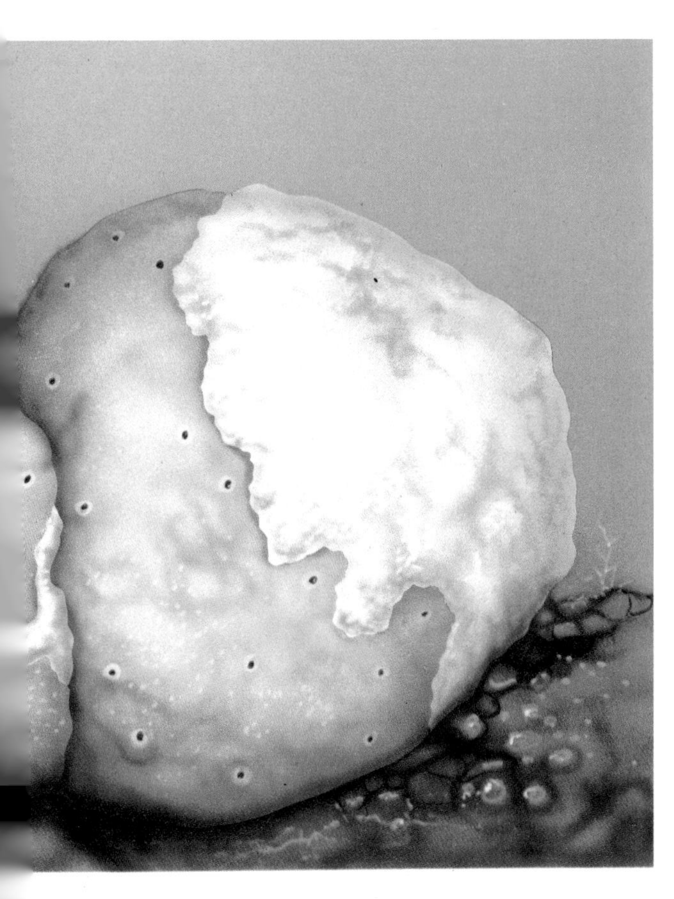

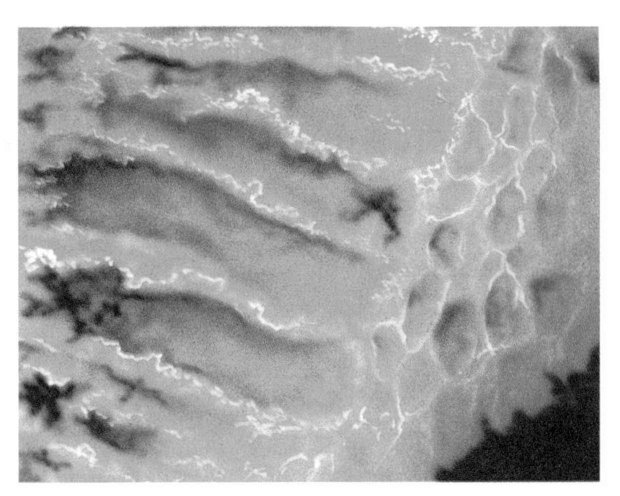

The complex interior structure of a Sycon sponge, magnified by an electron microscope.

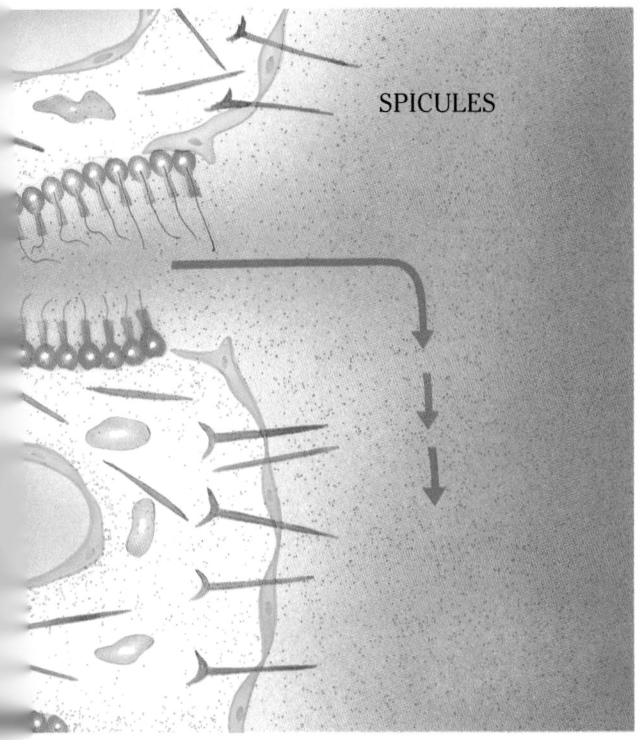

The sponge has ducts that bring in water from the outside and move it toward chambers lined with flagellated cells. This is the water filter system. Water enters through microscopic openings, or ostioles. These ostioles lead to incurrent ducts through which water passes until it reaches the flagellated chambers. The chambers are lined with small, hairy cells called choanae. Using their flagellum, the choanae catch food from the water. The filtered water then moves through excurrent canals and out the exit opening, or osculum.

Improving filtration

Sponges are different sizes and shapes. Sometimes they can even look like vegetables or colored rocks.

Environmental conditions, such as ocean currents and sedimentation, influence the external shape of the sponge. In some cases, sponges of the same species can look completely different because of their living conditions.

Sponges grow in three main shapes. Some sponges form thin, crusty layers. This happens if they live on a hard surface in an area beaten by waves.

Glass and tube-shaped sponges grow straight upward as the result of living on a deep horizontal surface that has large amounts of sediment.

Sponges that grow upright on the sea floor can create interesting formations.

Some sponges take the shape of the surface they live on.

Sponges can grow in rocky corners.

They grow this way so that the sediment on the bottom will not block their pores.

Some sponges have the shape of a tree with smooth branches. These sponges grow broadside to a steady water current to catch more food.

Some sponges grow like tree branches. This enables them to catch large amounts of food.

EATING AND BREATHING WHILE FILTERING WATER

Efficient filters

The amount of water a sponge filters depends on many factors — such as its size, number, osculum diameter, and sea current strength. For example, a small Leuconia sponge 4 inches (10 cm) high and 0.4 inches (1 cm) in diameter contains almost 2,250,000 chambers that can pump 24 quarts (22.5 liters) of water a day. The volume of water filtered daily by a sponge is usually enormous, and it can retain between 68 percent and 99 percent of the food particles that pass through its filtering system. On the other hand, the sponge does not have specialized respiratory organs. Instead, it breathes by taking oxygen directly from the water circulating through its interior. Some sponges can absorb more than 75 percent of the oxygen in the water.

Some sponges have incurrent pores grouped into specialized areas. This allows greater filtration capacity.

Sponges sometimes look very fragile and prefer to live in places with little light.

The Clathrina sponge changes shape as it contracts. It does this several times a day.

The sponge's skeleton

The sponge's skeleton is made of tiny calcareous or silicon spicules. This enables it to support a body full of openings and canals. Besides spicules, sponges can also have elastic protein fibers. Scientists use spicules to classify sponges. Each species has spicules of different forms and sizes. The various combinations of spicules and fibers lead to many kinds of textures. This explains how sponges can be elastic and voluminous or crusty and hard.

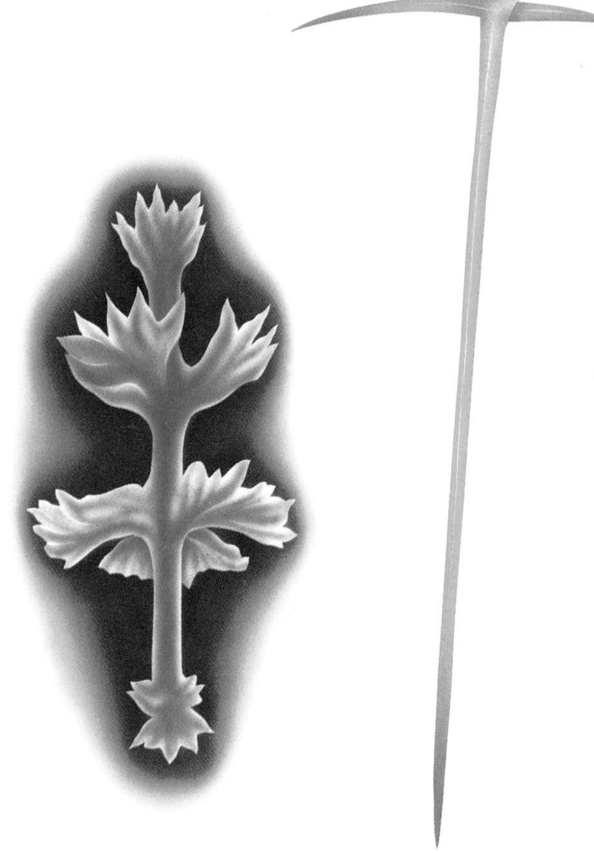

Spicules come in dozens of different shapes, some very unusual.

Did You Know...

that there are sponges 10 feet (3 m) long?

Although sponges vary in size, they are usually between 2-16 inches (5-40 cm) long. Some are smaller than a grain of rice, but there are also giant sponges more than 5 feet (1.5 m) in diameter that weigh 130 pounds (60 kg). Sponges of the Monoraphus family have even reached 10 feet (3 m) in diameter.

Growing speed varies according to the species. Sponges that live in fresh water and near the coastline usually grow more rapidly than others. The illustration below shows various shapes of sponge skeletons.

ANCESTORS OF THE SPONGE

The first sponges

Sponges spent many years of evolution perfecting the shapes they have today. Sponges that lived over 600 million years ago were simple organisms. The number and distribution of openings and chambers differed from those of today's sponges. However, the number of species increased a little at a time until about 400 million years ago. By that time, an example of each main sponge group existed.

Guettardia is one of the most common sponge fossils. It lived almost fifty million years ago.

This Chaetetes fossil shows the layerlike form of its internal skeleton.

A very long history

Sponges are among the oldest known animals. Remains have been discovered that show sponges lived in the sea almost one billion years ago.

Some fossils are abundant. In some cases, the silicone spicules are so numerous that they form layers or "rocks" which can be more than 330 feet (100 m) thick. After sponges with spicules not joined to the body die, these spicules fall and accumulate unevenly. In sponges with the spicules attached to the body, entire skeletons are preserved.

This entire Syphonia fossil has been preserved because its spicules were joined to its body.

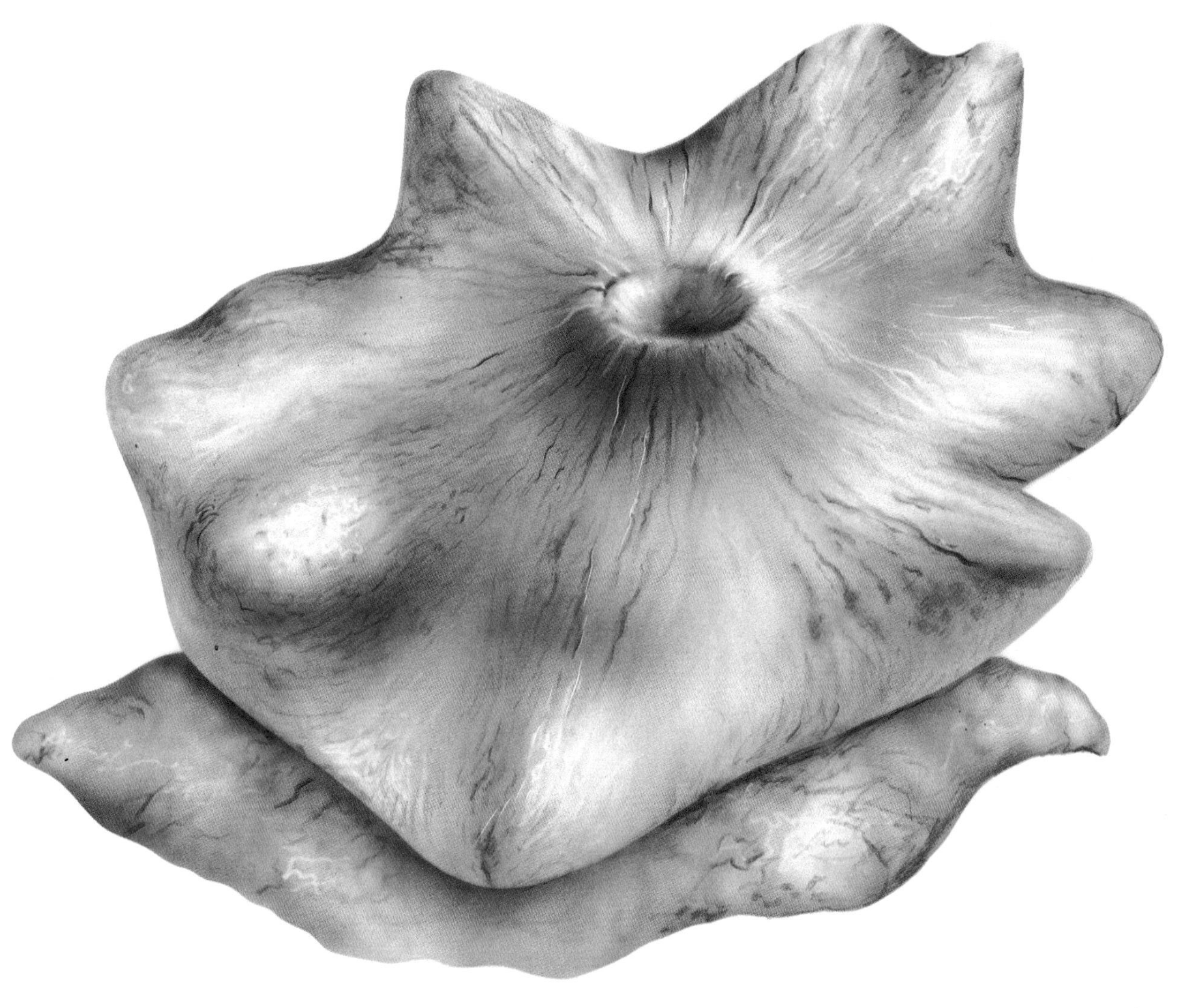

Did You Know...

that a broken sponge can rebuild itself?

If a piece of sponge is cut and its cells separated with a filter, the cells will take the form of an amoeba and begin to move around until they meet other cells.

The movements then stop and the cells form small groups that begin to grow. If the cells form a big mass, each cell then seems to remember the functions it had when it was still a part of the original sponge. The cells begin to reorganize and reconstruct a new, very small sponge that is ready to feed and grow.

THE LIFE OF THE SPONGE

Protection from enemies

Sponges do not have special sense organs, and they cannot move, but they are not defenseless.

Many sponges have sharp barbs, and some can produce irritating substances. This is why some crab species use sponges as a defense system and for camouflage. The sponge benefits by getting from one place to another because the crab carries it everywhere.

This Suberites sponge lives on a hermit crab's shell.

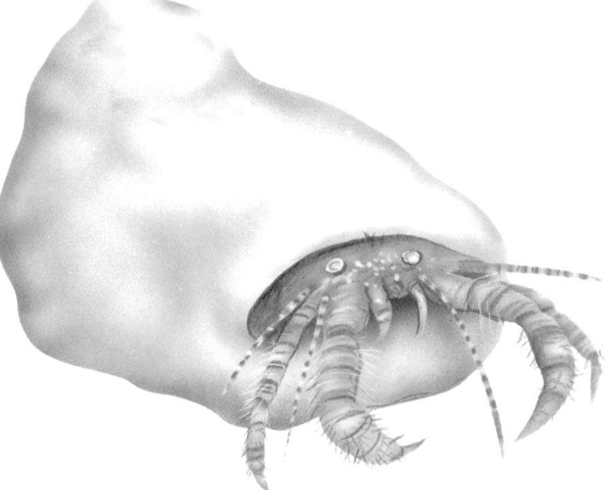

Peltodoris is a mollusk that feeds only on the tissue of the Petrosia sponge on which it lives.

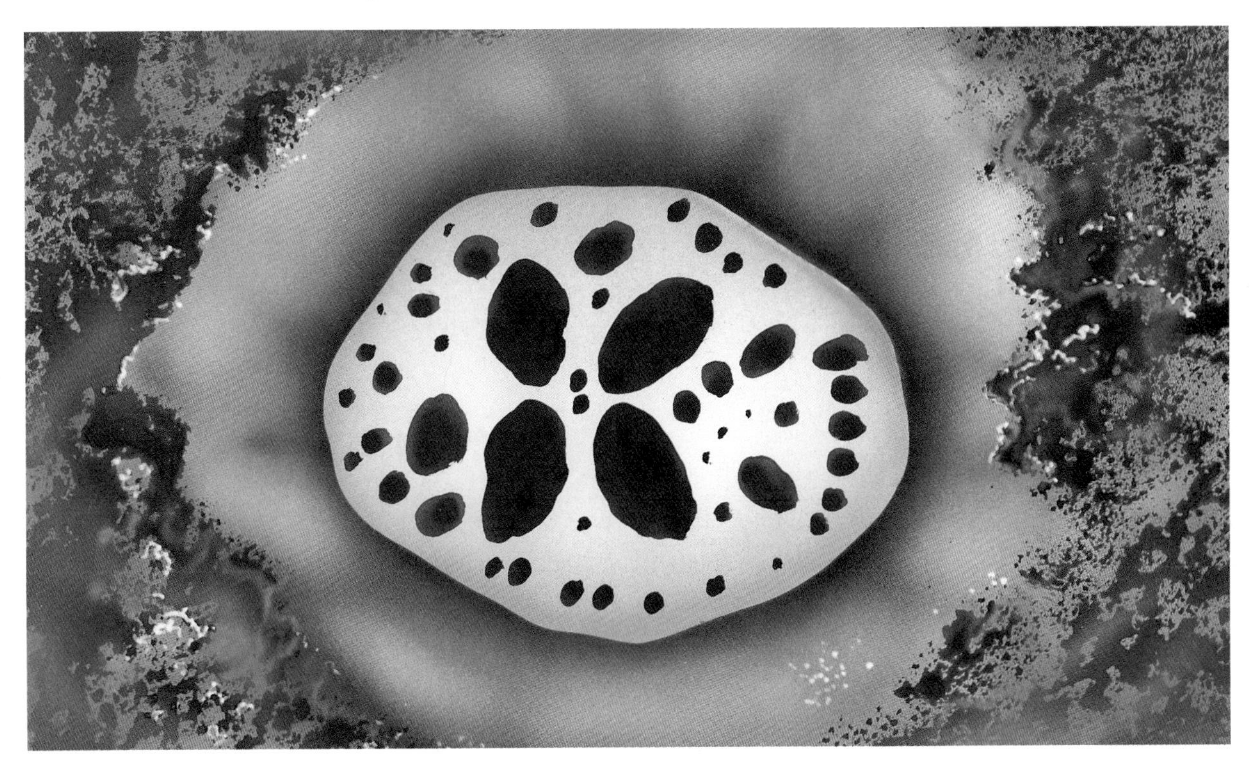

Did You Know...

that bath sponges are sponge skeletons?

The early Egyptians from the Mediterranean region were the first humans to discover the use of sponges for personal bathing.

These bath sponges, which are very popular today, are actually horny skeletons with spongy, netlike fibers that make it elastic and flexible.

One of the most common bath sponges from the Mediterranean is *Spongia officinalis,* which is round, springy, and bendable. It is light brown inside and dark gray or black outside.

Life inside a sponge

The sponge's structure is one of chambers and ducts — a good place for small animals to use as a temporary or permanent home. If you cut a sponge through the middle and examine it, you will see that it is like a tiny universe full of minuscule organisms of every kind. In Venus' flower-basket sponges, a pair of tiny shrimp can enter through the openings. Over time, the shrimp grow in size and eventually become so big they are unable to get out.

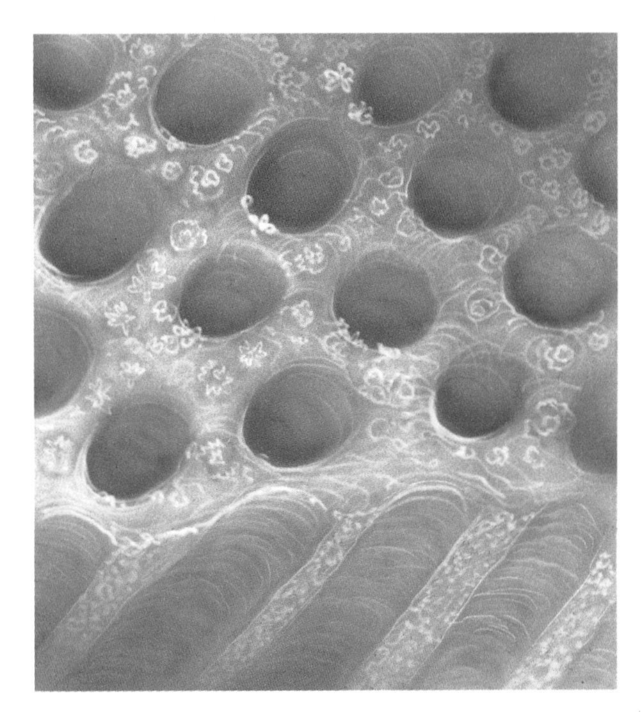

A sponge skeleton, seen under an electron microscope, looks ideal as a home for tiny animals.

This Dysidea sponge is a refuge for small organisms like these tiny polyps with tentacles.

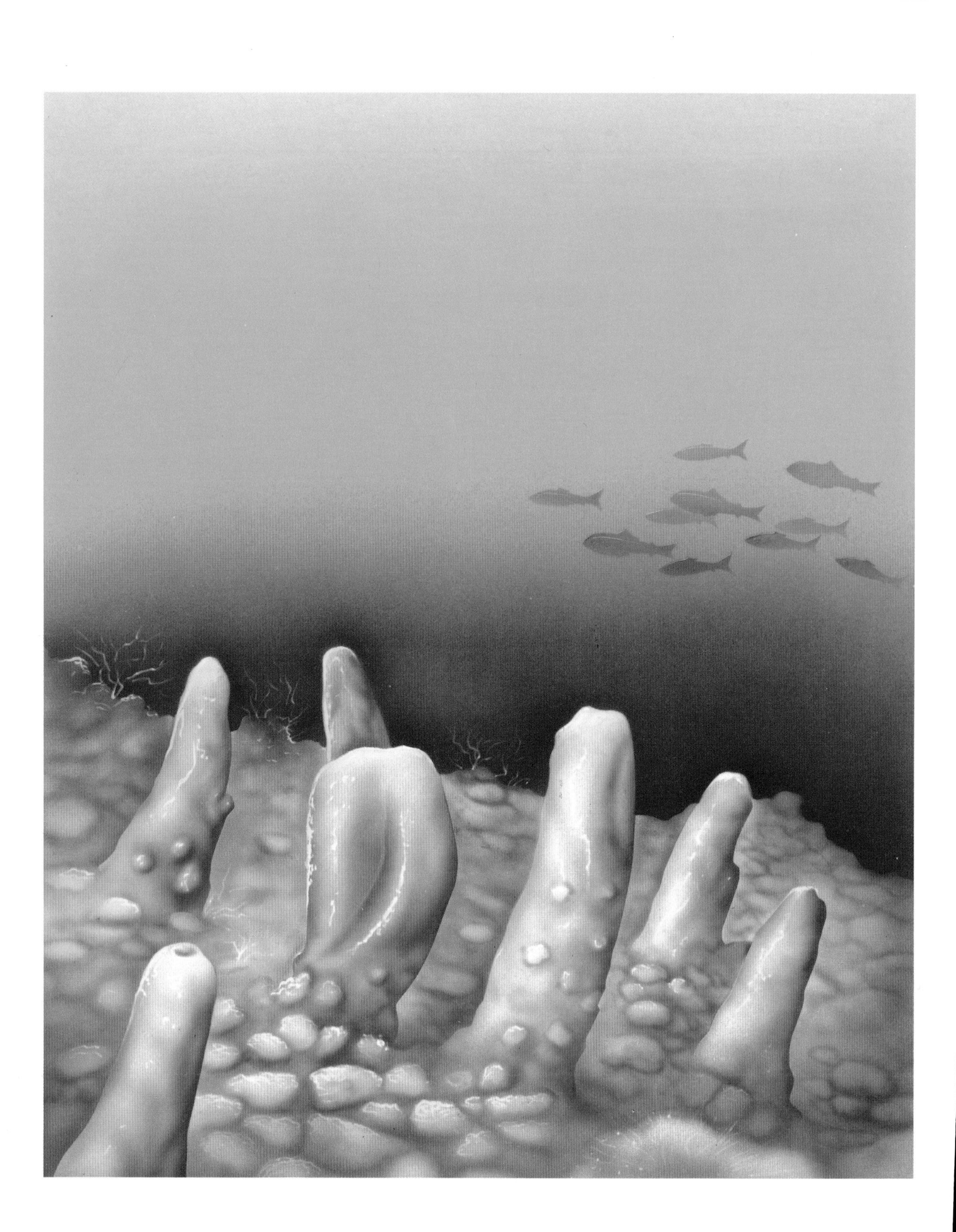

APPENDIX TO

SPONGES

Filters of the Sea

SPONGE SECRETS

▼ **An unappetizing dish.** Few animals eat the sponge because of its spiky skeleton and the poisonous substances it can produce. Its main enemies are some types of fish, starfish, and mollusks.

Self-restoration. Sponges can regenerate lost body parts. Fishermen who collect sponges leave pieces on the sea floor to grow for future collection.

Cup-shaped sponges. Roman soldiers used sponges for drinking wine during sea voyages instead of the heavy metal goblets of the time.

The sponge business. Humans still collect huge quantities of sponges for commerce. Cuba, for example, collects more than 18 tons (20 tonnes) a year.

▼ **Submarine warfare.** A submerged rock covered with sponges is really a battlefield. Different species of sponges fight ferociously with each other and with other organisms. They compete for living space and produce toxic substances to use in combat.

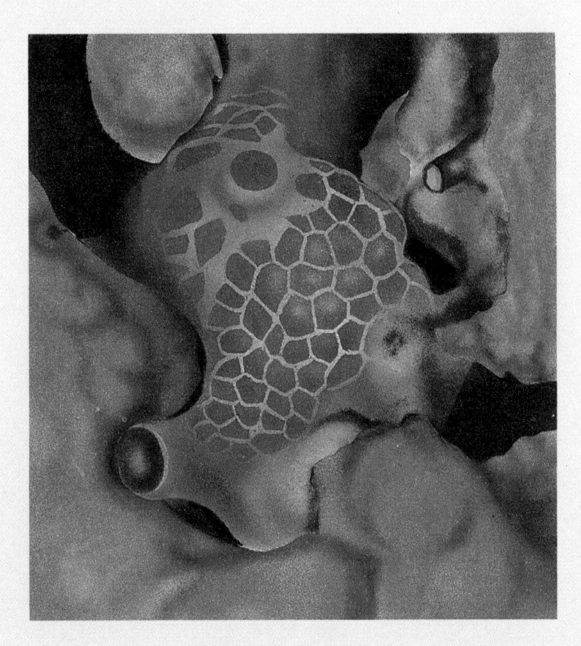

▸ **Destructive sponges.** Some sponges drill into and destroy the surfaces on which they live. These sponges are very dangerous for oyster beds or coral reefs because they can completely eliminate the inhabitants.

1. How many species of sponges are known to exist?
a) About 1,000.
b) About 10,000.
c) About 47,000.

2. The types of sponges and their order of classification are:
a) asconoid, syconoid, and leuconoid.
b) asconoid, syphonoid, and crustaceous.
c) ramifoid, syconoid, and calcaric.

3. Sponges filter sea water:
a) to stay humid.
b) to clean body openings.
c) to breathe and capture food.

4. Sponge cells that have a whiplike hair are called:
a) pores.
b) choanae.
c) spicules.
d) flagellae.

5. The types of sponges with the most efficient filter system are:
a) asconoids.
b) leuconoids.
c) syconoids.

6. The main exit area for water in the sponge is called:
a) the ostiole.
b) a pore.
c) the osculum.

The answers to SPONGE SECRETS questions are on page 32.

GLOSSARY

abundant: plentiful; having a large amount.

abyss: a bottomless area or space; the deepest parts of the oceans.

accumulate: to gather or collect something.

afferent ducts: ducts that carry something toward or into part of an animal.

amoeba: microscopic one-celled organism commonly found living in water. It can move around and easily change its flexible shape.

ancestors: previous generations of a family or species.

camouflage: a way of disguising something or someone to make it look like its surroundings.

capacity: ability to do something; the amount that can be used for storage or holding.

cavity: a hollow space inside something.

chamber: a small room; an enclosed space within something.

choana: a cell with a flagellum partly enclosed by a collar made from sticky membrane.

complex: complicated; made up of many different parts.

contract *(v)*: to pull in and make smaller or shorter.

current: a flowing mass of air or water.

duct: a tube through which a liquid or gas flows.

efferent ducts: ducts that carry something away from or out of part of an animal.

elastic: capable of recovering size and shape after stretching.

enable: to make possible, practical, or easy.

environment: the surroundings in which plants, animals, and other organisms live.

evolution: changing or developing gradually from one form to another. Over time, all living things evolve to survive in their changing environments, or they may become extinct.

expel: to push out or get rid of.

exterior: the outer surface.

external: outside of something.

ferocious: savage; brutal; fierce.

filter *(n)*: a porous mass through which a gas or liquid is passed in order to separate, or strain out, material it may contain.

flagellum (*pl* flagellae): long, whiplike thread that some single-celled organisms have.

flexible: able to bend or move with ease.

fossils: the remains of plants or animals from an earlier time period that are often found in rock or in Earth's crust.

fragile: easily broken or destroyed.

horizontal: in a position that is level with the ground.

incurrent: referring to a structure that allows current to flow in.

intercept: to stop something from moving any farther.

irritating: to bother or annoy; to cause soreness.

marine: of or related to the sea.

minuscule: very small.

nutritious: foods that help growth and development.

ovule: a small egg.

perfecting: to improve; to bring to final form; to make perfect.

pore: a tiny opening, especially in a plant or animal.

reconstruct: to reassemble; to put together again.

regenerate: to renew or restore to original strength or health; to regrow.

reorganize: to organize something again, either in the same or a different way.

residue: remnant; remainder; something that is left behind.

respiratory: anything to do with breathing.

retain: to keep.

sediment: the matter that settles to the bottom of a liquid.

sedimentation: the process of forming or depositing sediment.

sense organs: special organs of an

animal that help it hear, see, feel, or balance.

species: animals or plants that are closely related and often similar in behavior and appearance. Members of the same species are capable of breeding together.

specific: particular details or traits, like color or shape.

spicule: a minute, slender, hard structure that, in large numbers, supports sponge tissue. Spicules can be made from compounds with the minerals calcium (also found in bone) or silicon (also found in sand), and may be bound with protein fibers.

symmetry: an exact matching of parts or sections on opposite sides of a dividing line or around a central point.

toxic: poisonous.

voluminous: having great volume or bulk.

ACTIVITIES

- Natural sponges are used for many tasks. Find some library books that explain how sponges are harvested from the ocean and prepared for sale. What parts of the world have commercial sponge fisheries? Does pollution threaten the world's sponge supplies? What can people do about this?

- The different patterns of natural sponges can make beautiful craft designs. To make a pattern, use all or part of a sponge from a craft or art store or a store that sells house paint. Dip the sponge into poster paints, watercolors, or acrylics, squeeze out the excess, and carefully press the sponge on paper or cardboard. Use just one color or print several colors after the first color layer dries. Try printing darker colors over lighter ones. Use the paper to wrap presents. Gift boxes can also be printed this way for a special present! Make a background for an original picture, or use sponge prints to make a unique frame for your favorite piece of art.

MORE BOOKS TO READ

Aquatic Life. (Time Life)
Colors of the Sea series. Eric Ethan and Marie Bearanger (Gareth Stevens)
Coral Reefs. Alberto De Larramendi Ruis (Childrens Press)
Deep-Sea Vents: Worlds Without Sun. John F. Waters (Dutton)
Exploring Saltwater Habitats. Sue Smith (Mondo Publishing)
Great Barrier Reef. Martin J. Gutnik and Natalie Browne-Gutnik (Raintree/Steck-Vaughn)
Interesting Invertebrates. Elaine Landau (Franklin Watts)
Seashore Walk. (Running Press)
Simple Animals. Linda Losito, et al (Facts on File)
Small Sea Creatures. Jason Cooper (Rourke Corporation)
Sponges Are Skeletons. Barbara J. Esbensen (HarperCollins)

VIDEOS

Life and Death on the Great Barrier Reef. (Columbia Tristar Home Video)
Life on the Reef. (Dimension Media)
Marine Invertebrates. (Encyclopædia Britannica Educational Corporation)
Sponges, Anemones, Corals and Flatworms. (Educational Images)

PLACES TO VISIT

Seattle Aquarium
Pier 59, Waterfront Park
Seattle, WA 98010

Sydney Aquarium
Darling Harbour
Sydney, NSW, Australia

Auckland Institute and Museum
Auckland, New Zealand

Aquarium du Quebec
1675 Avenue des Hotels
Sainte-Foy, Quebec
G1W 4S3

Sea World on the Gold Coast
Sea World Drive Spit
Surfers Paradise
Queensland, Australia
4217

Aquarium of the Americas
Woldenberg Riverfront Park
New Orleans, LA 70130

Vancouver Aquarium
Stanley Park
West Georgia Avenue
Vancouver, British Columbia V6B 3X8

INDEX

Answers to SPONGE SECRETS questions:

1. **b**
2. **a**
3. **c**
4. **b**
5. **b**
6. **c**